ISAAC ASIMOV'S
Library of the Universe

Piloted Space Flights

by Isaac Asimov

Gareth Stevens Publishing
Milwaukee

Library of Congress Cataloging-in-Publication Data

Asimov, Isaac, 1920-
  Piloted space flights.

  (Isaac Asimov's library of the universe)
  Bibliography: p.
  Includes index.
  Summary: A brief history of man's experience in space as he circled the Earth, landed on the moon, and sent spaceships farther away than humans can safely go.
  1. Manned space flight—Juvenile literature. [1. Manned space flight] I. Title. II. Series: Asimov, Isaac, 1920-   . Library of the universe.
TL793.A823   1989   629.45'009   89-11303
ISBN 1-55532-371-5

A Gareth Stevens Children's Books edition
Edited, designed, and produced by
Gareth Stevens, Inc.
RiverCenter Building, Suite 201
1555 North RiverCenter Drive
Milwaukee, Wisconsin 53212, USA

Cover photography courtesy of NASA

Project editor: Mark Sachner
Series design: Laurie Shock
Book design: Kate Kriege
Research editor: Kathleen Weisfeld Barrilleaux
Picture research: Matthew Groshek
Technical advisers and consulting editors: Julian Baum and Francis Reddy

Printed in the United States of America

  2 3 4 5 6 7 8 9 95 94 93 92 91 90

CONTENTS

Nowadays, we have seen planets up close, all the way to distant Uranus and Neptune. We have mapped Venus through its clouds. We have seen dead volcanoes on Mars and live ones on Io, one of Jupiter's satellites. We have detected strange objects no one knew anything about till recently: quasars, pulsars, black holes. We have studied stars not only by the light they give out, but by other kinds of radiation: infrared, ultraviolet, x-rays, radio waves. We have even detected tiny particles called neutrinos that are given off by the stars.

Most wonderful of all, however, is that human beings have themselves ventured out into space. This is something that science fiction writers and artists have dreamed of for a long time, and it has finally come true. Human beings have not only circled Earth in orbit many times and returned safely; they have even stood on another world! In this book, I will tell you about putting people into space.

Isaac Asimov

The First Steps

Rockets have been around for centuries. The Chinese, for example, were using them over 700 years ago. And yet it has only been a few decades since humans first figured out how to get a rocket out of Earth's atmosphere and into space.

During World War II, Wernher von Braun, a German, developed rockets that could travel hundreds of miles. After the war, rockets went higher and higher. In 1957, the Soviet Union used a rocket to put a satellite into orbit.

If a satellite is big enough, it can carry a human being. On April 12, 1961, a Soviet rocket ship carried Yuri Gagarin once around the world and back safely. He was the first person in space.

The first US citizen in space was Alan B. Shepard. He made a "suborbital" flight — right up and down — on May 5, 1961. He was followed on August 6 by another Soviet, Gherman Titov, who completed 17 orbits of Earth, and by US space pioneer John H. Glenn, who circled Earth three times on February 20, 1962.

In the Soviet Union, these space travelers are called cosmonauts. In the United States, they are called astronauts.

The first American in space?
This chimpanzee, named Ham, flew a
suborbital flight before any human!

Inset, opposite: the first human in space,
Soviet cosmonaut Yuri Gagarin.

Opposite: A helicopter lifts Alan Shepard
out of the Atlantic Ocean after his
successful Mercury flight.

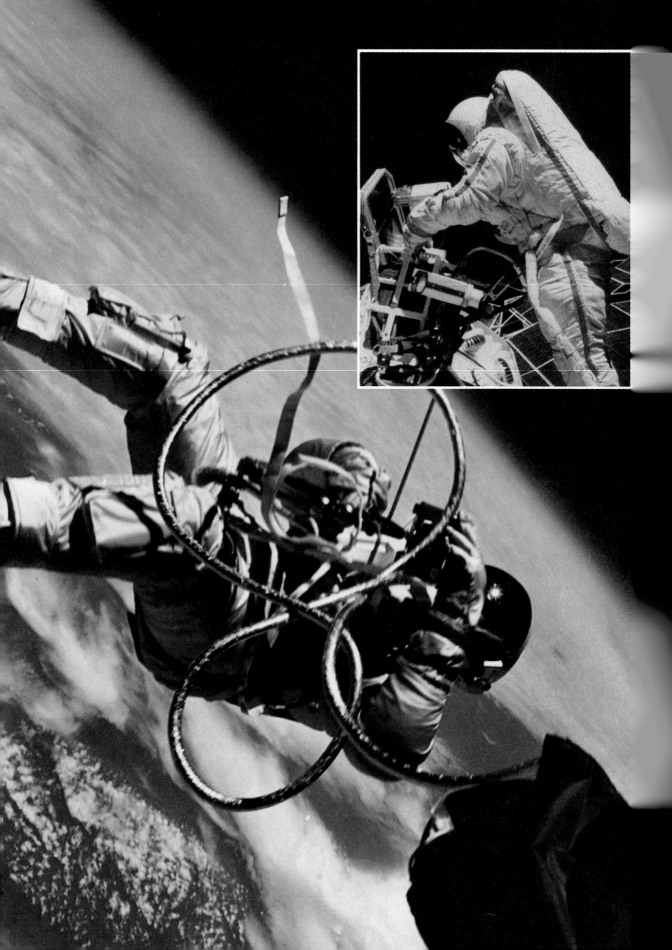

Walking in Space

Of course, it didn't take long for people to want to do more than stay inside a ship. On March 18, 1965, cosmonaut Alexei Leonov worked himself out of a Soviet ship that was in orbit around Earth. He was tied to the ship by a tether, and for 10 minutes he looked at the stars and at the Earth beneath him. It was the first "space walk."

On June 3, astronaut Edward H. White took a space walk from a US rocket ship. He stayed out in space for 21 minutes. Naturally, he had to wear a spacesuit while out there, one that supplied oxygen and controlled temperatures.

Do the Space Walk! Inset, opposite: In 1984 Soviet cosmonaut Svetlana Savitskaya, the first woman to go into space for a second time, also became the first woman to complete a space walk. Opposite: Gemini 4 astronaut Edward White takes a walk outside his capsule.

Below: Spacesuits give astronauts more than air. They must protect against tiny meteoroids, shade the harsh rays of the Sun, and maintain a constant temperature. NASA first tests new designs underwater, as shown in this picture.

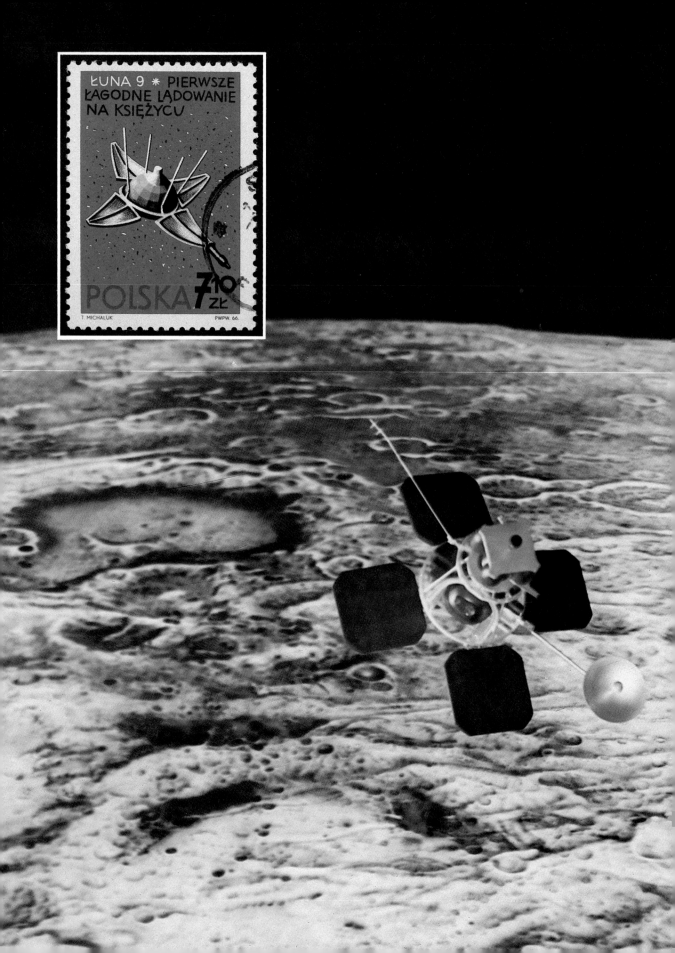

Taking Aim at the Moon

US President John F. Kennedy wanted US citizens on the Moon before the 1960s were over. But it had to be done a step at a time.

The US tested two-man ships in the Gemini program and three-man ships in the Apollo program. They had to work out rendezvous and docking, so two ships could approach each other and lock together. The first docking took place in March 1966 between Gemini 8 and an unpiloted craft. The US also sent rocket ships closer and closer to the Moon, and they sent one ship around the Moon and back, skimming just above the lunar surface.

The Soviet Union was interested in the Moon, too. In fact, they had even landed an unpiloted craft on the Moon's surface in September 1959. But the United States had set its sights on putting people on the Moon, and in 1969 the US was ready for the big step.

Opposite: Unmanned Lunar Orbiter scanned the Moon for possible landing sites. Inset: In 1966, a Soviet Luna probe (shown on a Polish stamp) made the first soft lunar landing.

Below: Apollo 12 astronauts landed about 600 feet (180 m) from Surveyor 3, which had set down on the Moon two years earlier.

Taking the Giant Leap

Finally, the spaceship Apollo 11 was ready to try landing on the Moon. On board were three astronauts — Neil Armstrong, Edwin E. "Buzz" Aldrin, and Michael Collins. Apollo 11 was launched on July 16, 1969.

Four days later it went into orbit around the Moon. As the lunar lander Eagle descended to the Moon's surface with Armstrong and Aldrin aboard, Collins remained in orbit aboard the command ship. On July 20, 1969, at 4:18 P.M. (Eastern Time), Eagle touched down on lunar soil. And at 10:56 P.M., Neil Armstrong became the first person in history to set foot on the Moon. As he stepped down, millions of people listening in on radio and TV heard him say, "That's one small step for man, one giant leap for mankind."

Life out there? — let's take a look!

People are always excited by the possibility of alien life. We're sure there's no life on the Moon. Our instruments tell us there's no life on Mars, though we should probably look for ourselves. Jupiter has a satellite, Europa, covered by a huge glacier. Might there be liquid water under the glacier — and life within it? Saturn has a satellite, Titan, with a thick atmosphere. Might there be some form of life on its surface? Let's take a look some day!

Opposite: Buzz Aldrin adjusts experiments during Apollo 11's brief visit to the lunar surface.

Below: President Nixon congratulates the crew of Apollo 11. They were isolated after their return to protect against alien germs.

Reusable Craft

In the early days of space flight, a rocket ship could be used only once. This seemed wasteful. Both the United States and the Soviet Union have developed a spaceship that can go up, return, and be used again, over and over. This kind of ship is called a space shuttle.

The first US space shuttle, Columbia, was launched on April 12, 1981 — 20 years to the day after Gagarin had made his first flight.

On November 15, 1988, the Soviet Union launched its first shuttle, the unpiloted Buran (a Russian word for "snowstorm" or "blizzard"). One of the tasks of a shuttle is to ferry construction units into space for building a space station.

Opposite: The US space shuttle Discovery lifts off from Cape Canaveral.

Inset, opposite: Tests with early designs like this HL-10 Lifting Body helped shape the US space shuttle.

Below: An artist imagines the Soviet shuttle Buran streaking into orbit. Inset: The European Space Agency is considering making a mini-shuttle named Hermes.

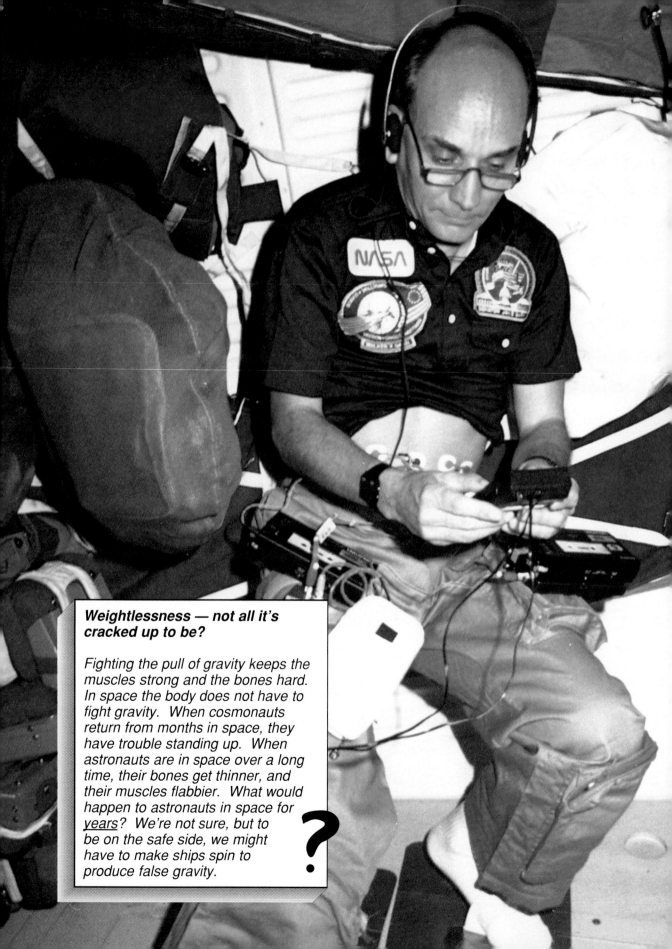

Weightlessness — not all it's cracked up to be?

Fighting the pull of gravity keeps the muscles strong and the bones hard. In space the body does not have to fight gravity. When cosmonauts return from months in space, they have trouble standing up. When astronauts are in space over a long time, their bones get thinner, and their muscles flabbier. What would happen to astronauts in space for <u>years</u>? We're not sure, but to be on the safe side, we might have to make ships spin to produce false gravity.

?

Taking the Short Hop

The Soviet Union has a space station, the Mir, in orbit at this very moment. Cosmonauts are brought to it and stay there for up to a year at a time. They are then brought back and replaced with others.

The US is planning a station where relays of astronauts can live and build space structures and new kinds of spaceships designed for short hops.

Assembled at the space station, these ships won't have to push through the atmosphere as they struggle to break free of Earth's full gravity. They'll carry people from the space station to where they are putting together a structure, and then back.

Opposite: US Senator Jake Garn tests his body's responses to weightlessness.

Top right: an artist's concept of the completed Soviet Mir space station.

Bottom right: An earlier Soviet space station, Salyut 7, held many cosmonauts. Although Salyut was small compared to Mir, unpiloted resupply ships brought oxygen, food, and equipment so that cosmonauts could make extended stays.

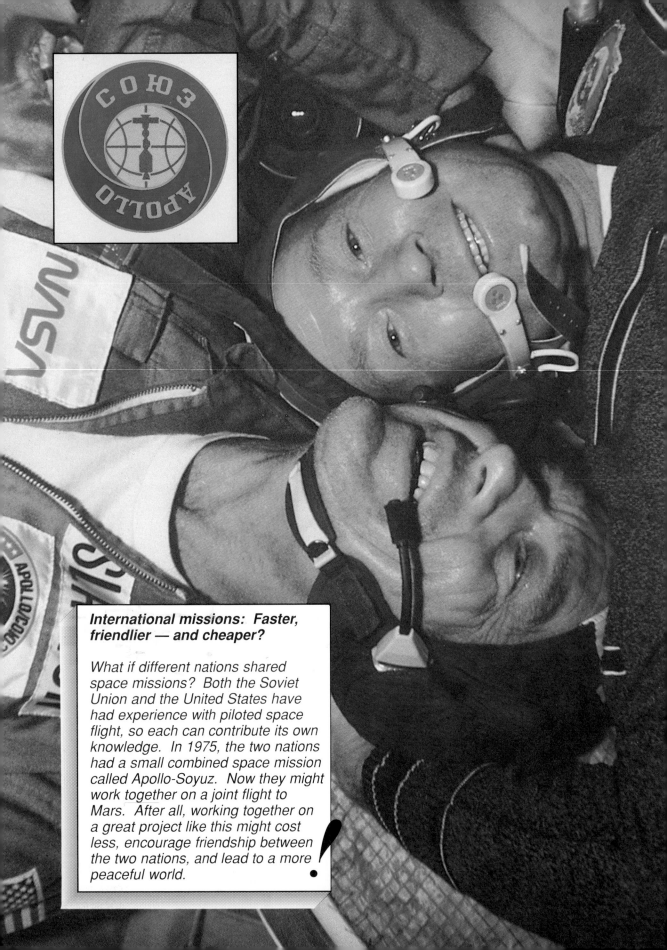

International missions: Faster, friendlier — and cheaper?

What if different nations shared space missions? Both the Soviet Union and the United States have had experience with piloted space flight, so each can contribute its own knowledge. In 1975, the two nations had a small combined space mission called Apollo-Soyuz. Now they might work together on a joint flight to Mars. After all, working together on a great project like this might cost less, encourage friendship between the two nations, and lead to a more peaceful world.

Putting Together a Crew

The first astronauts were men who had been to military test-pilot school and studied engineering or science. They had to understand complicated machinery, and because the first spaceships were quite small, they also had to be of medium height.

Valentina Tereshkova, the first woman in space.

Later on, civilian test pilots were chosen. In time, even civilians without piloting experience were chosen, too. Those chosen eventually included taller people. And finally, crews included civilians from the ranks of scientists, politicians, and teachers.

The selection of Soviet crews followed a similar pattern, although the Soviet Union led the way in putting women into space. The first woman to orbit Earth, Valentina Tereshkova, went up on June 16, 1963, while the first female US astronaut, Sally Ride, went up 20 years later, on June 18, 1983.

Opposite: Soviet cosmonauts and US astronauts on the same space mission? It happened in 1975 with the Apollo-Soyuz Test Project. Inset: an Apollo-Soyuz patch.

The varied crew of the US space shuttle Challenger included teacher Christa McAuliffe. All seven lost their lives when their craft exploded in 1986.

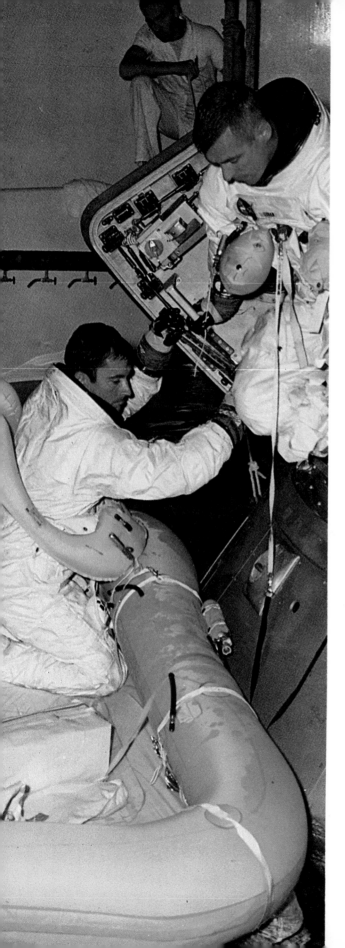

Training a Crew

Astronauts must understand how the controls of a spaceship work as well as the science and engineering behind them. They must also be tested physically and psychologically for the special conditions of space flight.

After all, astronauts live in small rooms for days and weeks. They experience high gravity when they take off and zero gravity in space. To help prepare astronauts for high gravity, their training includes whirling speedily around and around. For low gravity, they train underwater or dive in airplanes. Astronauts must also go through the motions of space flight in make-believe spaceships.

Left: Apollo astronauts practice an ocean recovery procedure.

Below: Astronauts train for missions in mockups of the US space shuttle.

Getting the feel of zero-g: Astronaut candidates experience weightlessness in a NASA airplane. The plane dives steeply toward the ground, creating weightlessness in a brief "free fall."

Facing the Problems

How well do people hold up in space? Shuttle flights and trips to the Moon usually take only a week or so, and people have managed that quite well. But to go beyond the Moon means that people will have to stay in their spaceships much longer.

How would humans fare on a one- or two-year trip to Mars and back? We don't know for sure, although the Soviets have successfully kept people in space for up to a whole year at a time.

And what about the spaceships themselves? After all, a ship must be able to support human beings for a long time. Oxygen, food, and water have to be supplied; carbon dioxide has to be removed from the ship's "atmosphere"; wastes have to be recycled.

Then, too, people living so close together for months at a time may have to work at not becoming irritated with each other!

No hands! This astronaut is about to grab that forkload of food floating before him.

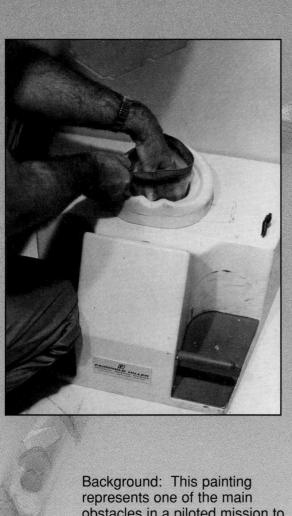

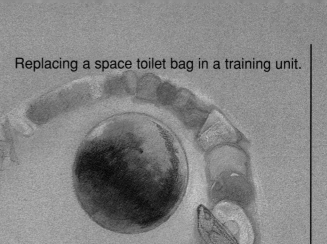
Replacing a space toilet bag in a training unit.

Would you believe . . .
chicken stew with
vegetables? Freeze-
dried, of course.

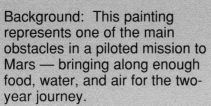

Background: This painting
represents one of the main
obstacles in a piloted mission to
Mars — bringing along enough
food, water, and air for the two-
year journey.

A Skylab food
tray. It even
warmed the food!

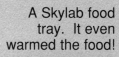

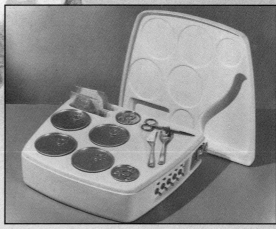

21

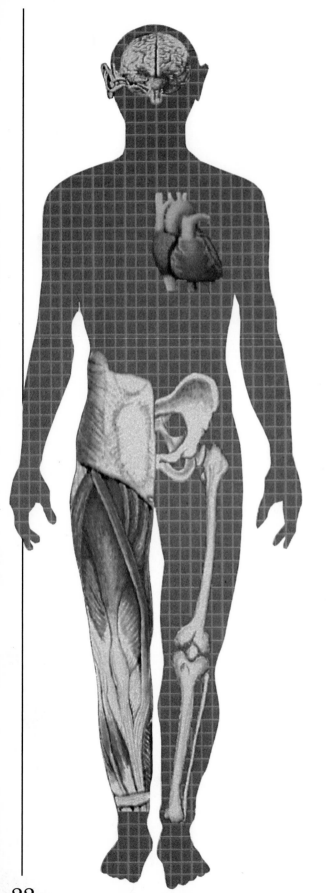

Our Future Neighborhood?

Flying to the Moon and back takes only a few days. But the other planets are much farther away than the Moon. At its closest to us, Venus is about two hundred times as far away as the Moon. Mars is 140 times as far away. Venus and Mercury are far too hot for human beings to land on, so that makes Mars our next target.

Any trip to Mars in the near future will take at least half a year. And, of course, after the astronauts have stayed on Mars for a while, they will have to come back — and that means at least another half a year. The trip will be a long, expensive, and dangerous one, but both the United States and the Soviet Union are thinking of trying it.

This illustration shows some areas of the human body that would be affected by long-term exposure to weightlessness. Muscles weaken when they no longer fight against gravity, the heart changes size, and the balance system of the inner ear may be affected. Even the very framework upon which we are built — the skeleton — begins to break down.

Astronauts are greeted by a tumultuous parade after their safe return to Earth. Will future space voyagers become instant celebrities upon their return from distant planets?

Space settlements — mobile homes to Mars?

We might try for a Mars flight now, but there may be better chances in the future. Once we erect a space station, we might use Moon material to build huge settlements in orbit where thousands of people might live. These "space people" should be much more used to space and space flight. It might be <u>they</u> who will make the long flight, since their settlements will be so much like spaceships, and they'll be used to low gravity and life-support systems.

?

An artist's concept of a mining operation drilling into the ruddy surface of Mars.

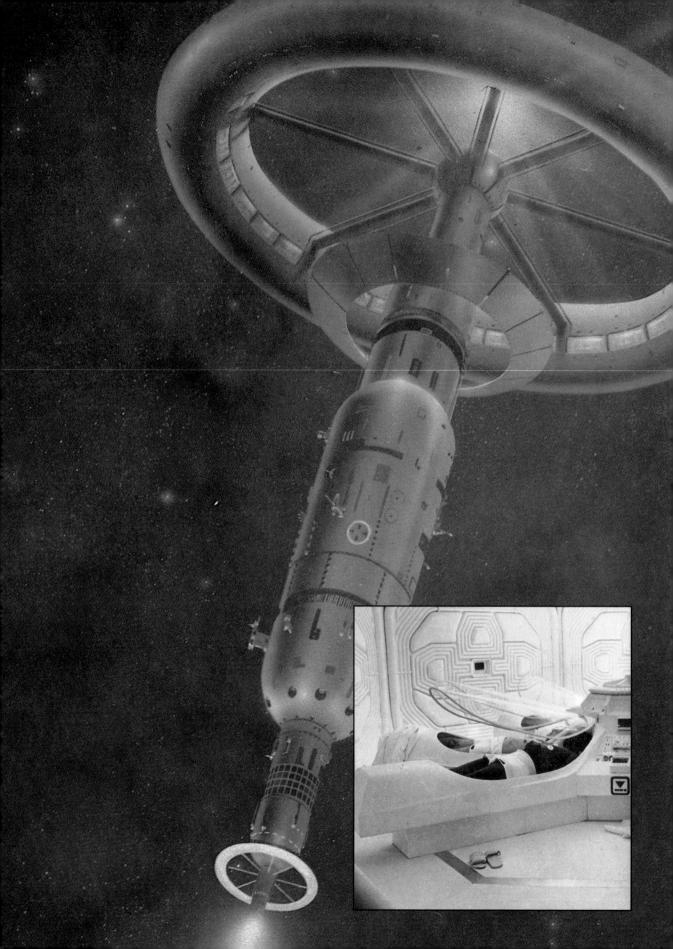

To Mars — And Beyond!

A trip to Mars and back would be quite a challenge. But Mars is part of the <u>inner</u> Solar system. Beyond it are the asteroids; then the giant planets Jupiter, Saturn, Uranus, and Neptune; then little Pluto and the distant comets.

The spacecraft Voyager 2 has visited the outer Solar system. It has gone to Jupiter, then Saturn, then Uranus, then Neptune, and sent back important information from each. Launched on August 20, 1977, Voyager took 12 years to reach Neptune.

That's a long time in the short life span of humans on Earth. But Voyager had no human beings aboard. Can astronauts make long trips like that? It doesn't seem so right now, but some day we'll have larger, faster, and better ships. Then? Maybe.

Opposite: A space settlement blasts away from Earth to explore a promising solar system. The ship rotates around its long axis, and occupants in the doughnut-shaped portion live with the normal sensation of gravity.

Inset, opposite: No need to for everyone to stay awake during the long ride to another planet. Many crew members could hibernate on the trip, awakened only when the vessel reaches its destination (as shown in this scene from the movie *Alien*).

Below: Many creatures, like the hermit crab, carry their homes around with them. Perhaps we could learn from their example in our quest to journey beyond the Solar system.

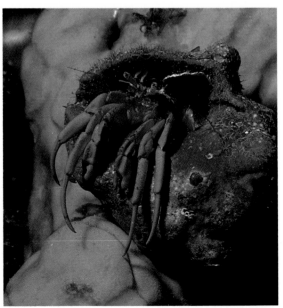

On to the stars — don't wait up!

Even the nearest star is about 7,000 times as far away as Pluto, the farthest known planet in our Solar system. Maybe the best way to visit the stars some day is for large settlements simply to cut loose and become independent little worlds, drifting through space with their loads of humanity for thousands of generations. They wouldn't feel lost or lonely because they would be taking their home with them. And what a view they'd be getting of the Universe!

Who Needs Space Pilots?

Voyager 2 did a marvelous job with no people aboard. Without human beings, a spaceship costs far less to operate. It doesn't need life-support systems. If something goes wrong with it, the results are not nearly as tragic as when the Soviet Soyuz 11 lost its cabin pressure on June 30, 1971, killing all three cosmonauts aboard, or when the US shuttle Challenger exploded on January 28, 1986, killing all seven crew members.

And with new telescopes and other instruments, we can see nearly to the edge of the known Universe. So if we can learn more about space from telescopes orbiting Earth than from a piloted ship on a dangerous mission to Mars, why send people to Mars?

There are many reasons for sending people into space. But there seems to be one fact we cannot escape: Human beings <u>want</u> to go out into space. It's more exciting and more dramatic. And let's face it, what would you rather watch bounding around on another world — a human being or a robot?

A future Soviet Mars mission will use a balloon to send a robot skimming over the planet's surface. It will set down each evening so the probe can take soil samples.

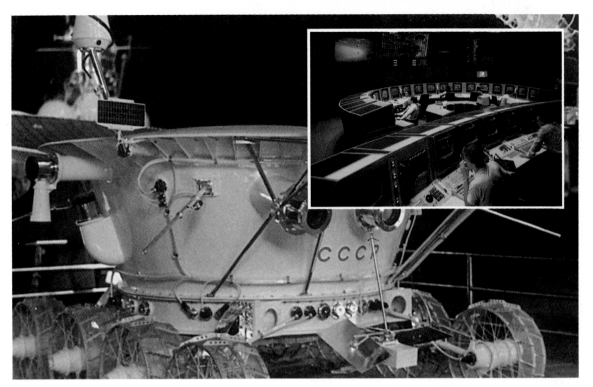

Above: The Soviet Lunakhod robot roamed the lunar surface, but was controlled from Earth. Inset: European Space Agency (ESA) control center for the flyby of Halley's Comet.

Below: The Viking orbiter, which reached Mars in 1976, gave us our most detailed view of Mars. Its lander craft beamed back photos from the Martian surface.

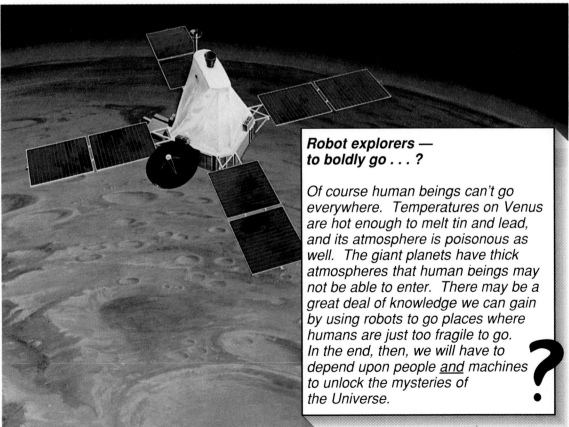

Robot explorers — to boldly go . . . ?

Of course human beings can't go everywhere. Temperatures on Venus are hot enough to melt tin and lead, and its atmosphere is poisonous as well. The giant planets have thick atmospheres that human beings may not be able to enter. There may be a great deal of knowledge we can gain by using robots to go places where humans are just too fragile to go. In the end, then, we will have to depend upon people <u>and</u> machines to unlock the mysteries of the Universe.

Fact File: "Space-on-Earth" Programs

Let's face it. It's one thing to read or to watch movies about piloted space flights. But plunging into the sights, sounds, and other sensations of space travel is quite another story. The simple fact is that not everyone can take part in space flight — at least not yet. There are so many requirements for a person who wants to go into space — good physical condition, education, and training are only three of them — that most of us will have to be content to read and watch from afar.

But if you've dreamed of being an astronaut, or at least of getting a feel for actually flying beyond Earth's atmosphere, don't lose hope. Many museums and space centers now give you a chance to experience some of the joys of space travel through camps, training simulators, and special shows.

Here is a list of some of these programs. Many of them began as programs for children but now appeal to grown-ups as well. One more thing to keep in mind: Some of the programs cost quite a bit of money.

Challenger Centers for Space Science Education

Houston, Texas
Washington, DC

More sites under consideration, perhaps as many as 50 one day, including some at schools and museums

• Simulated work places and space habitats, such as a mission control center and space station
• A variety of "missions" lasting from several hours to several days. Among existing or planned "missions": a trip to the Moon and a space station rendezvous with Halley's Comet
• Evening and weekend space science classes for children and grown-ups

Call **1-800-533-6310**, or write:
Challenger Center for Space
    Science Education
P.O. Box 90077
Washington, DC 20090

Aerospace Camp

Garden City, New York

Day camp featuring such activities as these:
• Building and launching model rockets
• Trying on space suits
• Simulated satellite repair work
• Field trips to Kennedy Airport and Grumman Aerospace Corporation

Call **1-516-222-1190**, or write:
Aerospace Camp
Cradle of Aviation Museum
Museum Lane, Mitchel Field
Garden City, New York 11530

Shuttle Camp

Alamogordo, New Mexico

Day camp or overnight programs featuring these activities:
• Building and launching model rockets
• Eating space food
• Handling actual space hardware, such as shuttle tiles and space helmets
• Underwater zero-gravity simulations
• Field trips to nearby White Sands Missile Range and Holloman Air Force Base

Call **1-800-545-4021**, or write:
Shuttle Camp
Space Center
Alamogordo, New Mexico 88311

Space Camps

Huntsville, Alabama
Titusville, Florida
Yawata, Japan (opening scheduled for spring 1990)

Programs of varying lengths with these features for children and grown-ups:
• "Space walk" in a Manned Maneuvering Unit (MMU) simulator
• Multi-axis simulator, which rotates in every direction to create the disorienting, sometimes wild, feel of navigating a craft
• Simulated shuttle launch, with group members taking turns acting as shuttle crew and mission control members
• "Space Habitat": housing that looks like a NASA space station
• Tours of nearby Alabama Space and Rocket Center and NASA Marshall Space Flight Center (Alabama) and NASA Kennedy Space Center (Florida)
• Japanese Space Camp: in addition to simulators and dorms similar to those in US Space Camps, features a museum, rides, and space theater

Call **1-800-63SPACE**, or write:
Alabama Space and Rocket
    Center
Space Camp Applications
Tranquility Base
Huntsville, Alabama 35807

A young space cadet tests her lunar legs in the one-sixth gravity chair, US Space Camp.

Michigan Space Camp

Jackson, Michigan

Week-long day camp featuring these activities:
• Building model rockets from scratch
• Visiting a planetarium
• Airport field trip
• Stargazing on a Friday night sleepover (optional for eighth graders)

Call **1-517-787-4425**, or write:
Michigan Space Camp
Attn. Glen Swanson
2111 Emmons Road
Jackson, Michigan 49201

Future Astronaut Training Program

Hutchinson, Kansas

Five-day program with such activities as these:
• Space survival training (eating and going to the bathroom in space)
• Flight and MMU simulator workouts
• A simulated shuttle launch

Call **1-316-662-2305**, or write:
Future Astronaut Training
    Program
Kansas Cosmosphere and
    Space Center
1100 North Plum Street
Hutchinson, Kansas 67501

Pacific Rim Spaceflight Academy

Portland, Oregon

Five-day "flight" programs with these features:
• Desert survival training
• Underwater training facility to demonstrate movement in zero gravity
• Helicopter and airplane rides

Call **1-503-222-2828**, or write:
Pacific Rim Spaceflight Academy
Oregon Museum of Science and
    Industry
4015 SW Canyon Road
Portland, Oregon 97221

More Books About Piloted Space Flights

Here are more books about piloted space flights. If you are interested in them, check your library or bookstore.

Colonizing the Planets and Stars. Asimov (Gareth Stevens)
How Do You Go to the Bathroom in Space? Pogue (TOR Books)
Out of the Cradle: Exploring the Frontiers Beyond Earth. Hartmann (Workman)
Sputnik to Space Shuttle: The Complete Story of Space Flight. Nicolson (Dodd, Mead)
The World's Space Programs. Asimov (Gareth Stevens)

Places to Visit

You can explore the realm of piloted space flights without leaving Earth. On pages 28-29 of this book, you can find several museums and space centers that offer "space-on-Earth" programs. Here are some additional places where you can find a variety of space exhibits.

Touch the Universe —
Manitoba Planetarium
Winnipeg, Manitoba

US Air Force Academy Planetarium
Colorado Springs, Colorado

Lockhart Planetarium
University of Manitoba
Winnipeg, Manitoba

Martin Marietta/Michoud Assembly Facility
New Orleans, Louisiana

Discovery Center of Science and Technology
Syracuse, New York

Margaret C. Woodson Planetarium/
    Horizons Unlimited — Supplementary
    Education Center
Salisbury, North Carolina

For More Information About Piloted Space Flights

Here are some places you can write to for more information about piloted space flights. Be sure to tell them exactly what you want to know about.

For information about planetary missions:
NASA Jet Propulsion Laboratory
Public Affairs 180-201
4800 Oak Grove Drive
Pasadena, California 91109

About rocketry and propulsion:
NASA Lewis Research Center
Educational Services Office
21000 Brookpark Road
Cleveland, Ohio 44135

NASA Marshall Space Flight Center
Educational Services Office
Huntsville, Alabama 35812

For a price list and information about space food (send a self-addressed, stamped envelope):
American Outdoor Products
1540 Charles Drive
Redding, California 96003

Glossary

alien: in this book, a being from some place other than Earth.

Apollo 11: the first piloted vehicle to land on the Moon. On July 20, 1969, Neil Armstrong became the first person to walk on the Moon.

Apollo-Soyuz: the last Apollo mission. Apollo and a Soviet Soyuz vehicle linked together in a joint US-USSR orbital mission in 1975. Three US astronauts met two Soviet cosmonauts in orbit to shake hands and conduct joint experiments.

asteroid: "star-like." The asteroids are very small planets made of rock or metal. There are thousands of them in our Solar system, and they mainly orbit the Sun in large numbers between Mars and Jupiter. But some show up elsewhere in the Solar system — some as meteoroids and some possibly as "captured" moons of planets such as Mars.

astronaut: a person who travels beyond Earth's atmosphere.

atmosphere: the gases that surround a planet, star, or moon.

carbon dioxide: along with nitrogen, one of the main gases that made up Earth's atmosphere early in the history of our planet. It is a heavy, colorless gas. Carbon dioxide is what gives soda its fizz, and when humans and other animals breathe, they exhale carbon dioxide.

civilian: someone not on active duty in a military, police, or firefighting force.

comet: an object made of ice, rock, and gas; has a vapor tail that may be seen when the comet's orbit brings it close to the Sun.

cosmonaut: an astronaut, especially one from the Soviet Union.

glacier: an enormous layer of ice formed from compacted snow, often itself carrying a layer of snow.

gravity: the force that causes objects like the Earth and Moon to be attracted to one another.

nuclear fusion: the smashing together of highly heated hydrogen atoms. This fusion of atoms creates helium and produces tremendous amounts of energy.

orbit: the path that one celestial object follows as it circles, or revolves, around another.

oxygen: the gas in Earth's atmosphere that makes human and animal life possible. Simple life forms changed carbon dioxide to oxygen as life evolved on Earth.

politician: a person who runs for or holds a government office.

relay: an electromagnetic device for remote or automatic control.

satellite: a smaller body orbiting a larger body. The Moon is the Earth's <u>natural</u> satellite. Sputnik 1 and 2 were Earth's first <u>artificial</u> satellites.

space shuttles: rocket ships that can be used over and over again, since they return to Earth after completing their missions.

space stations: artificial bodies in space in which humans can live and work, often for months.

space walk: an extravehicular (outside of the vehicle) venture made by an astronaut in space.

World War II: the second war fought by most of the principal nations of the world during the first half of the 20th century.

zero gravity: weightlessness.

Index

The publishers wish to thank the following for permission to reproduce copyright material: front cover, pp. 4 (large), 5, 6 (large), 8 (large), 9, 10, 11, 12 (inset), 14, 16 (both), 17 (lower), 18 (both), 19, 20, 21 (photographs), 23 (upper), 27 (bottom), courtesy of NASA; pp. 4 (inset), 6 (inset), 15 (both), 17 (upper), 27 (top, large), James Oberg Archive; p. 7, courtesy of McDonnell Douglas Astronautics Company; p. 8 (inset), from the collection of Ken Novak; p. 12 (large), courtesy of Rockwell International; p. 13 (lower left), © Paul Dimare; pp. 13 (upper right), 27 (inset, top), courtesy of European Space Agency (ESA); p. 21 (background art), © Rick Karpinski/DeWalt and Associates, 1989; p. 22, © Garret Moore, 1989; p. 23 (lower), artwork by Pat Rawlings, courtesy of NASA; p. 24 (large), © Ron Miller; p. 24 (inset), Photofest; p. 25, © Larry Lipsky, Tom Stack and Associates; p. 26, © Michael Carroll; p. 29, courtesy of US Space Camp/Academy.